8th International Congress on Sustainability Science and Engineering (ICOSSE'19)

Kuala Lumpur, Malaysia
1 – 3 July 2019

ISBN: 978-1-7138-0567-0

Printed from e-media with permission by:

Curran Associates, Inc.
57 Morehouse Lane
Red Hook, NY 12571

Some format issues inherent in the e-media version may also appear in this print version.

Printed with permission by Curran Associates, Inc. (2020)

AIChE
120 Wall Street, FL 23
New York, NY 10005-4020

Phone: (800) 242-4363
Fax: (203) 775-5177

www.aiche.org

Additional copies of this publication are available from:

Curran Associates, Inc.
57 Morehouse Lane
Red Hook, NY 12571 USA
Phone: 845-758-0400
Fax: 845-758-2633
Email: curran@proceedings.com
Web: www.proceedings.com

TABLE OF CONTENTS

SESSION 7 - SUSTAINABLE TECHNOLOGIES DEVELOPMENT II

Plenary Talk: Process Systems Engineering for Optimal Deployment of Negative Emissions Technologies

Conference: International Congress on Sustainability Science Engineering ICOSSE

Year: 2019

Proceeding: International Congress on Sustainability Science & Engineering (ICOSSE '19)

Group: Abstract Submissions

Session: Plenary Session I

Time: Monday, July 1, 2019 - 9:00am

Authors:

Tan, R., De La Salle University

Recent projections made in 2018 by the Intergovernmental Panel on Climate Change (IPCC) show that global greenhouse gas (GHG) emissions must taper off to zero by mid-century in order to ensure that mean atmospheric temperature rise is kept to a manageable level of about 1.5Â°C by 2100. Given this projection, carbon dioxide (CO_2) removal or negative emissions technologies (CDR/NETs) will be needed to offset growing GHG emissions from human activities, which are driven by economic and demographic growth trends. CDR/NETs extract CO_2 from the atmosphere and sequester the carbon through various mechanisms. Examples of CDR/NETs include so-called â€œnatural climate solutionsâ€ such as reforestation and soil carbon management, as well as novel techniques that include direct air capture (DAC), bioenergy with CO_2 capture and storage (BECCS), biochar application, indirect ocean capture (IOC), and enhanced weathering. Process systems engineering (PSE) methodologies will be essential to support decision-makers in the planning and design of CDR/NET systems as they are scaled up to significant levels in the coming two decades. In this talk, we discuss some recent PSE work on the use of mathematical programming, pinch analysis, and process graphs for optimizing CDR/NET systems based on BECCS, biochar application, IOC and enhanced weathering.

Combined Solar Thermal Treatment: Drying Characteristics and Kinetics of Oleochemical Sludge

Conference: International Congress on Sustainability Science Engineering ICOSSE

Year: 2019

Proceeding: International Congress on Sustainability Science & Engineering (ICOSSE '19)

Group: Abstract Submissions

Session: Session 1 - Waste handling, management and minimization I

Time: Monday, July 1, 2019 - 11:20am-11:45am

Authors:

Chong, C. H., Institution of Engineers Malaysia
Zailan, R. I., Heriot-Watt University Malaysia
Choo, C. M., Heriot-Watt University
Chok, V. S., KL-Kepong Oleomas Sdn. Bhd

The objectives of this study are to determine the combined thermal treatment drying characteristics and drying kinetics of palm oil olechemical sludge, impact of ultrasonic pre-treatment on drying duration, the effect of contained heat moisture removal using a solar dryer and the impact of combined drying technology. The drying characteristics for the combinations of all treatments including ultrasonic pre-treatment duration with different operating condition were investigated. When the drying temperatures increases from 70Â°C to 80Â°C, the range of drying rate increases from 1.02 â€“ 0.14 g H_2O/g DM (dry basis) for 70Â°C to 1.20 â€“ 0.06 g H_2O/g DM (dry basis). It was found that the changes on the drying characteristics are not obvious and there is no impact in increasing ultrasonic duration. Solar drying combined with thermal oven drying was conducted and the difference in terms of the effects of energy conservation were compared with thermal drying combined with ultrasonic pre-treatment and thermal drying alone. The period to reach equilibrium for solar drying combined with thermal drying and thermal drying alone is the same which can save two hours of energy for every 600 g of sludge. The cost of electricity can be saved by 22% if using solar drying combined with thermal drying compared with thermal drying alone.

The Kinetics and Thermodynamics of the Interaction between Phenol and Granular Activated Carbon Prepared from Waste Coconut Shell

Conference: International Congress on Sustainability Science Engineering ICOSSE

Year: 2019

Proceeding: International Congress on Sustainability Science & Engineering (ICOSSE '19)

Group: Abstract Submissions

Session: Session 1 - Waste handling, management and minimization I

Time: Monday, July 1, 2019 - 11:45am-12:10pm

Authors:

Wahid, S., The University of Trinidad & Tobago
Smith, J., University of the West Indies
Maharaj, R., The University of Trinidad & Tobago
Boodlal, D., The University of Trinidad & Tobago

Trinidad and Tobago (TT) is a highly industrialized an oil and gas based economy. Within recent times there has been serious health, safety and environmental concerns with regards to the level of the chemical contaminant phenol present in â€œproducedâ€ water generated within the oil and gas sector. Globally, the technique of Activated Carbon Adsorption of the phenolic compounds has proven to be a very successful however its utilization in TT has challenges as the material is expensive and has to be imported and this issue is further compounded due to a scarcity of US foreign exchange in TT in recent times. In an effort to develop an innovative solution to the problem, the possibility of converting coconut shell, a locally abundant waste material, to produce a microporous activated carbon is being explored. Success of this initiative will also provide an environmentally sustainable method for the disposal of large quantities of waste coconut shell which is also posing disposal challenges in TT at this time. The ultimate goal of successfully designing an adsorption column using microporous activated carbon produced locally entails producing the activated carbon, ensuring that its physical characteristics are within specifications and most importantly elucidating the thermodynamic and kinetic parameters as it relates to phenol adsorption after which the adsorption unit can be designed. Preliminary results have shown that the coconut shell is a good precursor and the optimum conditions for its production is being optimized. Work investigating the kinetic and thermodynamic models for adsorption is in progress and information derived will be used for the design of an adsorption unit.

Determining the Optimum Locations of a Locally-Managed Decentralized Solution for Conversion of Waste Plastic to Fuel in Developing Regions

Conference: International Congress on Sustainability Science Engineering ICOSSE

Year: 2019

Proceeding: International Congress on Sustainability Science & Engineering (ICOSSE '19)

Group: Abstract Submissions

Session: Session 1 - Waste handling, management and minimization I

Time: Monday, July 1, 2019 - 12:10pm-12:35pm

Authors:

Joshi, C., University of Kentucky College of Engineering
Seay, J. R., University of Kentucky

Due to the lack of sufficient capital, infrastructure and knowledge pertaining to hazards associated with waste accumulation, plastic waste management is a serious issue in rural regions of developing countries. As a result, in order to be effective, waste plastic management strategies must rely simply on the resources and technical knowledge of local communities to devise innovative solutions. This research proposes the implementation of a locally-managed decentralized circular economy (LMDCE) as a potential method for waste minimization by reusing and recycling waste plastic at its source of generation. A feasible LMDCE solution is conversion of waste plastic to fuel oil via thermal decomposition, utilizing appropriate technology in underdeveloped regions.

This research presents an initial supply chain optimization case study of a small-scale, low-cost, LMDCE based waste plastic to fuel oil process â€“ designed by the University of Kentucky â€“ in Jinja, Uganda. Utilizing MATLAB computational software, the optimum number, size, and locations of the technology for future implementation will be identified. The role of small entrepreneurs and waste pickers, amount of waste plastic completely removed from the ecosystem, and the sustainability of the proposed solution will further be discussed.

Modification of F-Doped TiO2 Nanorods with Rgo for Rhodamine B Photodegradation

Conference: International Congress on Sustainability Science Engineering ICOSSE

Year: 2019

Proceeding: International Congress on Sustainability Science & Engineering (ICOSSE '19)

Group: Abstract Submissions

Session: Session 1 - Waste handling, management and minimization I

Time: Monday, July 1, 2019 - 12:35pm-1:00pm

Authors:

Hidayat, R., Sebelas Maret University
Wahyuningsih, S., Faculty of Mathematics and Natural Sciences, Universitas Sebelas Maret
Ramelan, A. H., Electronic Materials and Energy Research Group, Universitas Sebelas Maret

TiO_2-based combination materials have been developed to enhance photodegradation performance. The combination of structural engineering modifications of nanorods, F atoms doping and composites with rGO was able to eliminate the problems that occur in photocatalyst material. The structural engineering of nanorods increases the surface area of TiO_2 so the interaction between dye molecules and photocatalyst also increases. Then, the presence of F atoms in the TiO_2 crystal lattice caused defects that allow the red shift in the UV-Vis wavelength range. The absorption shift was known from the characterization of the UV-Vis spectrophotometer. The rGO material is able to avoid recombination rate that often occur in photocatalytic processes. This material has a high surface area and good electronic conductivity so that can capture electrons (e^-) which are separate from holes (h^+). The formation of TiO_2-based combination materials was characterized by XRD, SAA, SEM and FTIR. Then, the photodegradation activity test was carried out on Rhodamine B with a time variation of 0, 30, 60, 90 and 240 minutes. The combination of modification of TiO_2-based photocatalyst material was able to enhance photodegradation performance compared to TiO_2 P25.

Invited Talk: The Integration of Safety, Health and environmental Indexes in Solvent Design Using Process Systems Engineering Tools

Conference: International Congress on Sustainability Science Engineering ICOSSE

Year: 2019

Proceeding: International Congress on Sustainability Science & Engineering (ICOSSE '19)

Group: Abstract Submissions

Session: Session 2 - Sustainable technologies development I

Time: Monday, July 1, 2019 - 1:50pm-2:15pm

Authors:

Chemmangattuvalappil, N. G., University of Nottingham - Malaysia

Computer-Aided Molecular Design (CAMD) is a powerful tool for designing solvents with targeted properties. The benefit of CAMD tools in solvent design is its ability to search for a vast number of conventional or novel potential solvents that exhibit the best performance based on the properties of interest. In most of the CAMD problems, the molecular physical and thermodynamic properties are often selected as the target properties during the design stage since they can be estimated through property prediction models. The generated set of molecules will then undergo performance analysis in order to determine the optimal solvent. The performance of the molecules is usually assessed using economic, safety and environmental indicators. While designing solvent molecules, it is important to incorporate safety, health and environmental (SHE) aspects along with target properties to ensure that the solvents do not cause hazardous conditions. The measurement of SHE parameters are based on the molecular properties that have impact on both of these aspects. The property involved in each parameter is assigned with an index score to represent the degree of potential hazards. These properties can be estimated through various property prediction methods. In an integrated design approach, SHE indexes can be estimated during the early stage of solvent design. A higher index value represents a higher hazard and impact level for the potential solvent. Hence, a solvent that satisfies the target properties and has the lowest penalty value may be a reasonable choice. However, the solvents with the best functionality may not necessarily exhibit a low penalty score. Therefore, a decision making must be made on the trade-off between the functionality of the solvent and its inherent safety and health level and potential environmental impacts. Various process systems engineering tools have been developed to incorporate the functionality and SHE targets simultaneously for solvent design.

Sustainability Analytics - the Critical Link between Sustainability and Business Innovation

Conference: International Congress on Sustainability Science Engineering ICOSSE

Year: 2019

Proceeding: International Congress on Sustainability Science & Engineering (ICOSSE '19)

Group: Abstract Submissions

Session: Session 2 - Sustainable technologies development I

Time: Monday, July 1, 2019 - 2:15pm-2:40pm

Authors:

Ramanan, R., Desert Research Institute

Advances in enterprise systems are making it feasible for corporations to track and transform performance to meet their corporate goals. Business analytics enable organizations to convert raw data into actionable insights to achieve their corporate performance objectives. In the sustainability context, governments and corporations are facing a plethora of ever-expanding management challenges related to issues spanning physical, financial, geo-political, social, and environmental concerns. The quadruple bottom line of sustainability comprises of profit, people, planet, and purpose â€“ a more recent enhancement. Sustainability analytics analyze the data and provide insights that help transform sustainability information into measures across the value-chain and life-cycle to accomplish their social responsibility and sustainability goals. The materiality of these seemingly non-economic impacts is the critical link between corporate sustainability and business strategies. Organizations must recognize the strong connections between competitive advantage and sustainability issues. This paper, based on the book, Introduction to Sustainability Analytics[i], introduces leaders to the untapped potential of sustainability analytics to determine which metrics are material, relevant and actionable to their business, and to prioritize sustainability initiatives.

[i] Ramanan, R., Introduction to Sustainability Analytics Ramanan, R., (Book), â€œIntroduction to Sustainability Analyticsâ€ ISBN: 9781498777056 and ISBN-10: 1498777058, CRC Press Taylor and Francis, July 2018

Optimization of Enhanced Weathering Networks with P-Graph and Pinch Analysis

Conference: International Congress on Sustainability Science Engineering ICOSSE

Year: 2019

Proceeding: International Congress on Sustainability Science & Engineering (ICOSSE '19)

Group: Abstract Submissions

Session: Session 2 - Sustainable technologies development I

Time: Monday, July 1, 2019 - 2:40pm-3:05pm

Authors:

Aviso, K., De La Salle University
Jain, S., Indian Institute of Technology Bombay
Bandyopadhyay, S., Indian Institute of Technology Bombay
Pimentel, J., Universidad Nacional de Colombia
Tan, R., De La Salle University
Friedler, F., Pazmany Peter Catholic University

Enhanced weathering is a negative emissions technology (NET) that is based on application of crushed CO_2-reactive rocks on agricultural or marginal lands. Such rocks react with atmospheric CO_2 and water from rainfall to form products that dissolve in runoff, and ultimately result in final sequestration of carbon in bodies of water. Scale-up of enhanced weathering as a commercial NET will require coordination of rock quarrying, crushing, and application in a supply chain network. Optimization models will be needed to facilitate planning of such enhanced weathering networks; a basic linear programming (LP) formulation has been developed by Tan and Aviso (2019). In this work, two approaches based on P-graph and pinch analysis are proposed as alternatives to the LP model. These methods are applied on an illustrative case study.

Homogeneous Catalytic Hydroesterification of Complex Alkenes - Built-in Base Ligands Pave the Way for Disruptive Technology

Conference: International Congress on Sustainability Science Engineering ICOSSE

Year: 2019

Proceeding: International Congress on Sustainability Science & Engineering (ICOSSE '19)

Group: Abstract Submissions

Session: Session 2 - Sustainable technologies development I

Time: Monday, July 1, 2019 - 3:05pm-3:30pm

Authors:

Kucmierczyk, P., Ruhr University of Bochum
Franke, R., Evonik
Beller, M., Leibniz Institute for Catalysis
Jackstell, R., Leibniz Institute for Catalysis
Sang, R., Leibniz Institute for Catalysis
Duehren, R., Leibniz Institute for Catalysis
Fridag, D., Evonik Performance Materials GmbH

The **hydroesterification** of highly **demanding alkenes** and industrially relevant bulk olefins (e.g. diisobuten) to corresponding esters has been studied. Herein a rationally designed Pd-catalyst based on a novel highly active class of **ligands (built-in base)** is used. Alkoxycarbonylation of a wide variety of alkenes in excellent yield is described. Initial steps towards the scale-up from laboratory to mini-plant-scale and validation of the process concept will be presented.

Results

The carbonylation of olefins is one of the most important homogeneously catalyzed reaction in industry. Within this group hydroesterification, also called Reppe reaction, represents a very promising and **environment- friendly one pot method** for obtaining carboxylic esters. Regularly, esters are produced by esterification of carboxylic acids with alcohols. In general, producing these carboxylic acids require multi-step processes, e.g. subsequent oxidation of alcohols or aldehydes obtained by hydroformylation.

In respect of green-chemistry, Pd-catalyzed hydroesterification of olefins has been studied quite extensively in the past several decades. However, the conversion of complex substrates like diisobuten to carboxylic esters is still challenging. Commonly used ligands, e.g. DTBPMB lead for complex olefins to significant lower yield and amounts of the unwanted byproduct ether. Fortunately, novel highly active ligands have been introduced for unprecedented conversion of **complex olefins**. [1,2] To obtain a framework of operation conditions, within constant product quality is guaranteed, **Design of Experiments** is used. Comparison of different separation technologies to **recycle this unique homogeneous catalyst**, lead to a sustainable process concept which is validated in a continuous operating mini-plant.[3]

[1] K. Dong, X. Fang, S. Gülak, R. Franke, A. Spannenberg, H. Neumann, R. Jackstell, M. Beller, *Nat. Commun.* **2017**, *8*,14117.

[2] R. Sang, P. Kucmierczyk, K. Dong, R. Franke, H. Neumann, R. Jackstell, M. Beller, *J. Am. Chem. Soc.* **2018**, *140*, 5217-5223

[3] P. Kucmierczyk, R. Sang, D. Fridag, M. Beller, R. Franke,**manuscript in preparation**

Coal Quality Evaluation System for Improving Thermal Power Plant Emission: A Malaysian Experience

Conference: International Congress on Sustainability Science Engineering ICOSSE

Year: 2019

Proceeding: International Congress on Sustainability Science & Engineering (ICOSSE '19)

Group: Abstract Submissions

Session: Session 2 - Sustainable technologies development I

Time: Monday, July 1, 2019 - 3:30pm-3:55pm

Authors:

Mohamad Ali, M. Z., TNB Fuel Services Sdn. Bhd.

This paper will present an overview of our experience in evaluating coal quality using a utility-grade software to ensure sustainable energy. The software is intended to evaluate expected emissions and fuel suitability for coal-fired power plants based on proposed fuel quality parameters. With the implementation of Incentive-Based Regulation (IBR) mechanism in Malaysia, it is imperative to maintain the most economical fuel cost, at the same time providing the best quality and environmental-friendly electricity to customers. It is intended with this software that it will increase the coal availability options to coal-fired power plants, hence minimizing the cost of coal procurement. The software will also assist in reducing coal-fired power plant emission in terms of predicting emission behavior during plant operation. This will also improve plant availability and maintain security of electricity supply to the National Grid System, while at the same time move a step closer to achieving a cleaner power generation.

Plenary Talk: Biomass Conversion to Biofuel

Conference: International Congress on Sustainability Science Engineering ICOSSE

Year: 2019

Proceeding: International Congress on Sustainability Science & Engineering (ICOSSE '19)

Group: Abstract Submissions

Session: Plenary Session II

Time: Tuesday, July 2, 2019 - 9:15am-9:40am

Authors:

Yusup, S., Universiti Teknologi PETRONAS

Energy is a key enabler of economy. Search for environmental friendly and greener fuel is increasing worldwide. Biofuel such as biodiesel, bioethanol, bio-paraffin which covers bio-gasoline, kerosene and green diesel, biogas and bio-hydrogen among others are the emerging biofuels. The talk focusses on biofuel production through hydrodynamic cavitation technology, catalytic biomass conversion and catalytic hydro-processing technologies of oleic acids to green diesel. A new strategy in producing synthetic diesel hydrocarbons from a mono-unsaturated fatty acid model compound such as oleic acid is highlighted. With 12 Principles of Green Chemistry and Green Engineering, the quest for greener, renewable and affordable bio-based fuel will be viable. A holistic approach through understanding the intrinsic nature of the feedstock, processes and products and the impact on human health and environment is the driver towards attaining sustainable agenda.

Planning and Scheduling Integration for Land Use Under Food-Energy-Water Nexus Considerations: A Case Study in Texas Edwards Region

Conference: International Congress on Sustainability Science Engineering ICOSSE

Year: 2019

Proceeding: International Congress on Sustainability Science & Engineering (ICOSSE '19)

Group: Abstract Submissions

Session: Session 3 - Food, energy and water nexus

Time: Tuesday, July 2, 2019 - 10:30am-10:55am

Authors:

Xiao, X., Institute of Process Engineering, Chinese Academy of Sciences
Nie, Y., University of Chinese Academy of Sciences
Avraamidou, S., Imperial College London
Pistikopoulos, E. N., Texas A&M Energy Institute, Texas A&M University
Li, J., The University of Manchester

Land use planning and scheduling under Food-Energy-Water Nexus (FEW-N) considerations is a complicated decision-making problem with resource competitions and conflicting objectives [1]. Systematic thinking based on FEW-N is necessary for modeling and optimization of land use systems [2]. However, challenges arise in making decisions while encountering conflicting objectives, multi-scale and multi-period problems, and multiple stakeholders [3]. To address these challenges, we developed a generic optimization-based land allocation approach, which provides i) a composite FEW-N metric to help solve the multi-objective optimization problem and carry out assessments [4, 5], and ii) a two-stage decomposition strategy to solve the multi-scale and multi-period planning and scheduling problem. The developed strategy was applied in an agricultural land use case study within the Texas Edwards Region. Computational results indicate that the approach can provide a comprehensive FEW-N metric to select strategies for optimal land allocation, limit stresses in the FEW-N, and achieve trade-off solutions for the multi-scale and multi-period FEW land use systems.

Reference:

[1] Tian, Hanqin, et al. "Optimizing resource use efficiencies in the foodâ€“energyâ€“water nexus for sustainable agriculture: from conceptual model to decision support system." Current Opinion in Environmental Sustainability 33 (2018): 104-113.

[2] Daher, Bassel, et al. "Developing socio-techno-economic-political (STEP) solutions for addressing resource nexus hotspots." Sustainability 10.2 (2018): 512.

[3] Garcia, Daniel J., and Fengqi You. "The water-energy-food nexus and process systems engineering: a new focus." Computers & Chemical Engineering 91 (2016): 49-67.

[4] Avraamidou, Styliani, et al. "Towards a Quantitative Food-Energy-Water Nexus Metric to Facilitate Decision Making in Process Systems: A Case Study on a Dairy Production Plant." Computer Aided Chemical Engineering. Vol. 43. Elsevier, 2018. 391-396.

[5] Nie, Yaling, et al. "A Food-Energy-Water Nexus approach for land use optimization." Science of The Total Environment 659 (2019): 7-19.

Investigation of Circular Economy from Sustainability Point of View Using a Global Integrated Model

Conference: International Congress on Sustainability Science Engineering ICOSSE

Year: 2019

Proceeding: International Congress on Sustainability Science & Engineering (ICOSSE '19)

Group: Abstract Submissions

Session: Session 3 - Food, energy and water nexus

Time: Tuesday, July 2, 2019 - 10:55am-11:20am

Authors:

Hanumante, N., Monash University
Shastri, Y., IIT Bombay
Hoadley, A., Monash University

Research in the field of the circular economy is mainly focused on economic benefits [1,2]. Hence, its holistic assessment from a sustainability point of view, specifically addressing the ecological implications, is essential. This research gap is addressed here.

An integrated global socio-economic-ecological model originally developed by the US-EPA is selected for this work [3,4]. This is based on a food web model with multiple trophic levels, and human-essential industries including one producing consumer goods. The economic dimension is integrated using the macro-economic price setting model.

Simulations are carried out for different levels of consumption (2 to 10 times) over a period of 200 years. The industrial sector depends on agriculture for raw materials. The collapse of the agricultural sector because of over-exploitation is observed. After incorporating the circular economy in the form of an industry that recirculates the used consumer goods, with a constant circulation of 17%, this collapse is delayed by at least 20 years.

The maximum level of circulation, the delay before adoption and the rate of adoption define the pathways for the adoption of the circular economy. 810 scenarios, 90 pathways with 9 consumption levels, are simulated. Aggressive adoption of the circular economy delayed the collapse of the agricultural sector further, in several cases leading to sustainable behaviour throughout the simulation horizon of 200 years. However, for very aggressive adoption pathways, the open access grassland (which competes with the agricultural sector for nutrients) collapses because of high agricultural demand. The set of solutions that provides sustainable pathways form a sustainability corridor which becomes narrower with increasing consumption, thus reducing the operating space for the policymakers.

This work, thus, establishes the benefits of incorporation of the circular economy from sustainability point view and underlines the importance of the dynamics of adoption of the circular economy.

On the Optimization of Water-Energy Nexus in Shale Gas Production Under Price Uncertainties

Conference: International Congress on Sustainability Science Engineering ICOSSE

Year: 2019

Proceeding: International Congress on Sustainability Science & Engineering (ICOSSE '19)

Group: Abstract Submissions

Session: Session 3 - Food, energy and water nexus

Time: Tuesday, July 2, 2019 - 11:20am-11:45am

Authors:

Oke, D., University of the Witwatersrand
Mukherjee, R., Gas and Fuels Research Center, Texas A&M Engineering Experiment Station
Sengupta, D., Texas A&M Engineering Experiment Station
Majozi, T., University of the Witwatersrand
El-Halwagi, M., Texas A&M University

This study presents a robust framework for water-energy nexus optimization in shale gas production and distribution network under uncertainty. Sustainable design of the network is achieved through treatment of wastewater using thermal membrane distillation (TMD), whereby a detailed design model is incorporated within the network superstructure in order to account for the energy requirement of the treatment unit. The proposed model also takes into account the scheduling problem associated with production of shale gas, optimal shale gas production and distribution networks. There are various uncertainties associated with the network. Among different uncertain variables, uncertainties associated with price and demand are most crucial as they can impact the optimal configuration significantly. Incorporation of uncertainty in the model seeks to meet the demand of natural gas consumers while accounting for the uncertainties associated with the price of natural gas, natural gas liquids and electricity. In this work, uncertainties are modelled via randomly generated scenarios using beta distribution of the price, based on historical data. Three different scenarios are generated from past data. The stochastic model is applied to a case study through the maximization of total profit and comparison is made with the results from deterministic approach. Results from the three different scenarios show 14.47%, 10.98%, and 12.26% increase in the profit, respectively, compared to deterministic approach. Savings in the freshwater requirement for fracturing and in energy associated with water management amounted to 23.2% and 42.7%, respectively.

On the Development Resources and Waste Management Strategies in Industrial Clusters

Conference: International Congress on Sustainability Science Engineering ICOSSE

Year: 2019

Proceeding: International Congress on Sustainability Science & Engineering (ICOSSE '19)

Group: Abstract Submissions

Session: Session 3 - Food, energy and water nexus

Time: Tuesday, July 2, 2019 - 11:45am-12:10pm

Authors:

Shehab, S., Texas A&M University at Qatar
Ahmad, R., Texas A&M University at Qatar
Linke, P., Texas A&M University at Qatar
Al-Mohannadi, D., Texas A&M University at Qatar

The recent Intergovernmental Panel on Climate Change (IPCC) report on the impact of global warming states that global warming is likely to reach 1.5 °C between 2030 and 2052 if it continues to increase at the current rate. This puts a strain on industry and on hydrocarbon resource dependent economies, such as Qatar, to reduce to emissions while maintaining profitability and growth. Most emissions stem from stationary industrial sources, often in the form of industrial clusters (cities or parks). This poses a challenge to manage the multiple feedstocks, many conversion processes, products, intermediates, wastes, and emissions that flow within industrial clusters. Several approaches exist to reduce CO_2 emissions using renewable energy, energy efficiency or Carbon Capture Utilization and Sequestration (CCUS). Thus, the development of an integrated holistic approach to systematically screen through the many possible strategic options - of managing resources and waste emissions- is crucial for many countries with natural resource centric economies and as well as for energy companies. This work allows the integration of various resources such as material including intermediates and energy in an industrial cluster produce to value added products, while adhering to emission restrictions. The industrial cluster studied, consist of several plants, which are interconnected by an existing infrastructure that allows exchanging materials, power, and heat. The resource management tool will be able to assess a given industrial cluster integration potentials or can design a new grass-root industrial cluster. A case study will be solved to illustrate the method and investigate the impact of CO_2 reduction strategies on the design of industrial clusters.

Impact of Feedstock Attributes on the Performance of Processing in Cellulosic Ethanol Applications

Conference: International Congress on Sustainability Science Engineering ICOSSE

Year: 2019

Proceeding: International Congress on Sustainability Science & Engineering (ICOSSE '19)

Group: Abstract Submissions

Session: Session 4 - Biomass conversion and utilisation I

Time: Tuesday, July 2, 2019 - 1:30pm-1:55pm

Authors:

Thompson, D. N., Idaho National Laboratory
Hartley, D., Idaho National Laboratory

Feedstock supply system predominately utilized in the production of cellulosic biofuels, is typically designed around a single biomass type available within a small supply radius. While the limited supply area reduces transportation cost, there is little opportunity to buffer against the wide variability of physical and chemical properties encountered during processing and conversion. Feedstock variability is a major contributor to delays and disruptions in the preprocessing and conversion systems. Classical approaches to eliminating delays and disruptions utilize a sequential approach, which leads to solutioons that do not consider interactions across the system. To examine the impact of feedstock variability on processing and conversion performance, Discrete Event Simulation (DES) models were developed for preprocessing operations of a herbaceous feedstock supply system, including feedstock property-based equipment performance, equipment failure modes (biomass properties causing equipment failures), and the expected time to repair or correct these failures. Initial models were developed to represent a commercial scale biorefinery (focusing on preprocessing and infeed to conversion reactor), with failure and repair information based on industrial operational experience. Initial runs of the models resulted in a capacity utilization of 29.1% of design. Following this, conversion performance data were also included to integrate aspects of both chemical and physical performance into the model. The performance of the physical system was based on pilot-scale data, which produced similar results as the initial simulations and had a capacity utilization factor of 28.9%. Measured conversion yield, based on the feedstock variation, resulted in an average yield of 88.1% of design. This methodology provides a way to dynamically examine biorefinery operational performance in relation to feedstock variability. A new metric, termed â€œOperational Reliability,â€ is defined to capture operational and yield impacts and provide an overall picture of how well the system operates; for the system modeled, modeled operational reliability was 25.5%.

A Superstructure-Based Modeling Approach for Supply Chain Network Design Optimization of Biomass Conversion and Utilization

Conference: International Congress on Sustainability Science Engineering ICOSSE

Year: 2019

Proceeding: International Congress on Sustainability Science & Engineering (ICOSSE '19)

Group: Abstract Submissions

Session: Session 4 - Biomass conversion and utilisation I

Time: Tuesday, July 2, 2019 - 1:55pm-2:20pm

Authors:

Khor, C. S., Universiti Teknologi PETRONAS
Abdul Razik, A. H., Universiti Malaysia Pahang
Elkamel, A., Khalifa University

Supply chain network design for biomass conversion and utilization is one of the major areas with influence on biomass-related technological progress and commercialization activities. This paper contributes towards optimizing a biomass to products supply chain by considering multiple cost factors including biomass resource acquisition cost, production cost, and transportation cost as well as direct sales to meet market demands . A superstructure-based modeling approach provides alternatives of biomass processing routes towards an objective of maximizing annualized profit. The formulated model entails five echelons of biomass processing is implemented on a practical supply chain design case study that involves a biomass-based manufacturing company in southwestern Ontario, Canada intent on long-term business expansion and product portfolio improvement. The results obtained indicates that a profit on the order of CDN 22.62 million (Canadian dollar) per year can be expected to be achieved, which is mainly derived through the sales of biofiller, bioethanol, and byproducts from the milling and sieving plant. Importantly, the developed model demonstrates the applicability of such a model-based approach in offering insights on operational flexibility for attaining economic decision-making on biomass resource utilization and processing technology selection.

Continuous Saccharification of Pretreated Lignocellulose in Simulated Moving Bed Reactor

Conference: International Congress on Sustainability Science Engineering ICOSSE

Year: 2019

Proceeding: International Congress on Sustainability Science & Engineering (ICOSSE '19)

Group: Abstract Submissions

Session: Session 4 - Biomass conversion and utilisation I

Time: Tuesday, July 2, 2019 - 2:20pm-2:45pm

Authors:

Jayakumar, V. P., Indian Institute of Technology Madras
Jayaraman, G., Indian Institute of Technology Madras

Conversion of cellulose to glucose is a complex process and its kinetics is governed by a number of substrate and enzyme related factors. In theory, continuous removal of the glucose minimizes the product inhibition on the enzymes which in turn leads to higher conversion at the expense of less enzyme which can be achieved using Packed Bed Reactors (PBR). PBRs are also demonstrated to have higher productivity in terms glucose release and offer the flexibility for enzyme recycling.

Thorough understanding of the hydrolysis pattern in single column PBR is the key in developing a viable process strategy for continuous hydrolysis of lignocellulosic biomass - employing multiple up-flow PBRs, with periodic switching of enzyme/buffer feed, akin to a Simulated Moving Bed (SMB) process. In this study we employed a four column - four zone, SMB reactor (SMBR) with each column in a specific and separate stage of hydrolysis. A fresh column of biomass entering first zone will pass through each zone during each switching interval and at the end of fourth switching interval will get unpacked and subsequently packed with new biomass. At steady state operation, one fresh column of biomass will be added to the loop during every switching.

In SMBR, with switching intervals of 10 hours and enzyme loading of 20 $FPU/g_{biomass}$, we achieved ~38% conversion and glucose concentration of ~29 g/l. The cellulose conversion in SMB process was comparable to that of batch slurry experiments at similar conditions and 80% of that achieved in a single column PBR at the expense of 40 $FPU/g_{biomass}$. Moreover, the SMBR system led to more than two fold increase in throughput in comparison to batch processes. We intend to further improve the efficiency of SMBR by a model based process optimization and couple it to fermentation for production of value added metabolites.

Exploration of Rarely Researched Algal and Lignocellulosic Biomass to Produce Hydrocarbons and Fuel Additive

Conference: International Congress on Sustainability Science Engineering ICOSSE

Year: 2019

Proceeding: International Congress on Sustainability Science & Engineering (ICOSSE '19)

Group: Abstract Submissions

Session: Session 4 - Biomass conversion and utilisation I

Time: Tuesday, July 2, 2019 - 2:45pm-3:10pm

Authors:

Kishore, N., IIT Guwahati
Kawale, H., IIT Guwahati

The petroleum depletion over a period of time and the need of value added petrochemicals drives a force to develop fruitful technologies. The recent trends in renewable energy production are yet to qualify the energy needs. Thermal processes for biofuels from biomass results biofuels in form of solid, liquid and gas which can be utilized as the fuel and value added products. Among these technologies, pyrolysis is a cost effective technology on pilot scale as well as it can be deliver to form a bio-refinery. Authors conducted pyrolysis experiment at 550 °C temperature for the conversion of algal biomass and at 700 °C for conversion of lignocellulosic biomass to get bio-fuels and hydrocarbons. Both the sources of biomass are available at tremendous amount around the globe but scarcely researched. Main product obtained is pyrolytic bio-oil, the bio-oil from algal biomass resulted in the higher heating value (HHV) of 16.59 MJ/kg whereas bio-oil from lignocellulosic biomass resulted in HHV of 20.32 MJ/kg. Further the bio-oils are characterized using Fourier Transform Infrared Spectroscopy (FT-IR) and the analysis shows the no. of functional groups present in them. Gas Chromatography - Mass Spectroscopy (GC-MS) results in large number of hydrocarbons present in the bio-oils. Thus, these characterization techniques revealed the fuel potential of both the biomass to be used as an energy-crops for production of energy in a sustainable manner.

Theoretical Framework for Sustainable Organic Waste Collection for Large-Scale Biogas Production in Johannesburg, South Africa

Conference: International Congress on Sustainability Science Engineering ICOSSE

Year: 2019

Proceeding: International Congress on Sustainability Science & Engineering (ICOSSE '19)

Group: Abstract Submissions

Session: Session 4 - Biomass conversion and utilisation I

Time: Tuesday, July 2, 2019 - 3:10pm-3:35pm

Authors:

Awosusi, A., University of Pretoria
Heydenrych, M., University of Pretoria
Matambo, T., University of South Africa
Hildebrandt, D., University of South Africa

The production of biogas on a large-scale is problematic for three major reasons; 1) cost of transport, storage and processing of waste biomass, 2) carbon footprint as characteristic of the logistical demands of waste biomass collection and transport 3) net energy of biorefinery. A global economy platformed on energy and commodities as produced from biomass feedstocks, contrary to the unsustainable production from the fossil refining analogue, is not only a possibility but potentially viable. Despite advancement in biorefining systems, a major bottleneck is the effect of the aforementioned challenges on the sustainability and economy of scale of biogas production, hence the need for optimized models for more efficient centralized production platforms. With a focus on sources of organic wastes in the the Johannesburg metropolis and documented knowledge of their co-digestibility, this work capitalizes on trends in smart transport to describe sustainable collection system for organic waste feedstocks for biogas production. Sensitivity analysis using spatiality correlated greenhouse gas (GHG) emission indicators, energy demand, biomethane potential and other cogent variables was done towards achieving set sustainability objectives across pre-processing for anaerobic digestion of waste biomass feedstocks.

Sustainable Process Design for Biofuel Production Via Syngas Conversion Pathway

Conference: International Congress on Sustainability Science Engineering ICOSSE

Year: 2019

Proceeding: International Congress on Sustainability Science & Engineering (ICOSSE '19)

Group: Abstract Submissions

Session: Session 5 - Biomass conversion and utilisation II

Time: Tuesday, July 2, 2019 - 3:30pm-3:55pm

Authors:

Tan, E., National Renewable Energy Laboratory

The primary goal of a sustainability evaluation is to assess the impact of systems on areas sought to be protected and maintained over time, such as human well-being and ecosystems [1]. Sustainable process design to convert biomass to fuels and chemicals plays an integral part in the bioeconomy sustainability. It is recognized that the integration of sustainability in process design is core to the mission in developing renewable fuels and should be considered a best practice in biorefinery design. A myriad of approaches and methodologies for the design and evaluation of sustainable chemical processes have been reported. A plethora of indicators has been proposed to evaluate different dimensions of sustainability. Various sustainability dimensionality (such as economy and environment), efficiency (i.e., energy and mass), and categories (i.e., stand-alone indicators, impact categories) have been incorporated into the sustainability in biorefinery design. In this presentation, performance indicators are evaluated with the GREENSCOPE (Gauging Reaction Effectiveness for ENvironmental Sustainability with a multi-Objective Process Evaluator) methodology, which evaluates processes in four areas: Environment, Energy, Economics, and Efficiency. The method develops relative scores for indicators that allow comparisons across various technologies. This talk will explore the different methodologies for sustainability process design and provide an example application of the GREENSCOPE approach on biomass gasification to cellulosic hydrocarbon technology, as well as examples of learnings from these sustainability assessments and their impacts to support and inform on-going R&D efforts.

References:

1. Schaubroeck T, Rugani B. A Revision of What Life Cycle Sustainability Assessment Should

Entail: Towards Modeling the Net Impact on Human Well-Being. *J. Ind. Ecol.* 21(6), 1464â
€"1477 (2017).

Pyrolysis of Food Waste for Sustainable Production of Biofuels in Marginal Enviroments

Conference: International Congress on Sustainability Science Engineering ICOSSE

Year: 2019

Proceeding: International Congress on Sustainability Science & Engineering (ICOSSE '19)

Group: Abstract Submissions

Session: Session 5 - Biomass conversion and utilisation II

Time: Tuesday, July 2, 2019 - 3:55pm-4:20pm

Authors:

Makkawi, Y., American University of Sharjah
El Sayed, Y., American University of Sharjah
Hassan Pour, F., American University of Sharjah
Khan, M., American University of Sharjah
Hakawati, R., American University of Sharjah
Badrelzaman, M., American University of Sharjah

Recently, the International Energy Agency (IEA) reported that bioenergy would have the biggest growth in renewable resources between 2018 and 2023 (IEAâ€™s Renewables 2018). The great potential lies in the use of biofuels in industry, electricity generation, hydrogen energy and transportation sectors. The use of waste biomass and organic residues for biofuels is particularly attractive as it offers a solution to mitigate concern over land use for energy crops besides addressing environmental problems associated with contamination of waste and soil by landfill. In addition, biomass conversion to biofuels produces biochar as a by-product, hence expanding the benefit of biomass energy to countering land degradation and having enhancing properties for some soils, for example improved water retention in soils of aridic to torric moisture regimes or alleviating high bulk density in compacted, heavy clay soils.

This study presents experimental work on the pyrolysis of food waste (leftovers from various restaurants, hospitals, and hotels) in the United Arab Emirates. The experiments were carried out in a lab scale auger pyrolysis reactor with a throughput of 500 g/h. The results include, in addition to the waste chemical and physical characteristics, detailed analyses of the products, bio-oil and char, for potential applications in energy production and as soil amendment, respectively. The study concludes with a critical assessment of the food waste against other alternative organic matters that are available for pyrolysis in marginal environments.

Suitability of Biochar Produced from Banana Peels As a Source of Bioenergy

Conference: International Congress on Sustainability Science Engineering ICOSSE

Year: 2019

Proceeding: International Congress on Sustainability Science & Engineering (ICOSSE '19)

Group: Abstract Submissions

Session: Session 5 - Biomass conversion and utilisation II

Time: Tuesday, July 2, 2019 - 4:20pm-4:45pm

Authors:

Selvarajoo, A., Nottingham University Malaysia Campus
Hanson, S., University of Nottingham, Malaysia

Extensive studies have been carried out to produce useful product by converting biomass waste. One of the products that can be derived from biomass is biochar, also considered as solid biofuel. Biochar obtained from biomass can be considered as an alternative source of renewable energy. If properly harnessed, renewable energy from biomass can supply to the global energy demand. Pyrolysis is one of the thermal conversion methods that converts biomass to biochar without the presence of oxygen. In this work, biochar were produced from banana peels. Experimental work was conducted using an electric tubular furnace and a Thermogravimetric Analyzer (TGA). The effect of temperature was investigated together with heating rate and residence time. Results show that the biochar which has the highest energy content considering char yield and heating value was obtained at 325 °C. The char yield was 47 %wt and its gross heating value of 25.9 MJ kg^{-1}. The residence time was 30 minutes. The heating rate within the range investigated had very little effect on the char yield and heating value. The results highlighted the effect of pyrolysis parameters on the yield and heating value of banana peels biochar, which could be advantageous to be utilised as alternative solid biofuel.

Blend of Pre-Treated Lignocellulosic Biomass and Bentonite As Water-Based Drilling Fluid: Preparation, Characterization and Testing

Conference: International Congress on Sustainability Science Engineering ICOSSE

Year: 2019

Proceeding: International Congress on Sustainability Science & Engineering (ICOSSE '19)

Group: Abstract Submissions

Session: Session 5 - Biomass conversion and utilisation II

Time: Tuesday, July 2, 2019 - 4:45pm-5:10pm

Authors:

Adewole, J. K., International Maritime College
H. S. Al-Maharouky, R., International Maritime College
Najimu, M. O., University of South Carolina
S. Hakeem, A., King Fahd University of Petroleum & Minerals
Bakare, I.
Adewunmi, A., King Fahd University of Petroleum & Minerals

Products from lignocellulosic materials have the potential to be used as substitutes for (or complement to) some of the oilfield chemicals. This study reports the outcome of a preliminary study of their effects on the performance of water-based drilling fluids containing bentonite.

Methodology

Date pit was used as a representative lignocellulosic material. Pretreatment was performed by dispersing 4g of the pit in 20ml of concentrated HCl (38%) at 35°C under constant stirring of 500rpm for 5hours. The resulting sample was then used for drilling fluid preparation. Tap water (350ml), commercial bentonite (22.5g) and 1-20wt% of the pre-treated lignocellulose was used for the preparation. Filtration, density, and rheological properties of the formulations were evaluated using the API test procedure. Physico-chemical properties were also studied using TGA, SEM, XRD, size analyzer, and FTIR.

Results and Observations

Physico-chemical analysis of the pretreated additive showed that the chemical structure of the material is still intact. However, the pretreatment loosens up the structure of the material by distrupting its crystalline regions. In the case of filter cake removal with enzymes, the de-crystallization will increase the surface area and the porosity of the lignocellulose making it more available for attack by the enzyme. The sample containing 20wt% of the lignocellulose material has filter cake thickness of 0.167cm (compare to 0.192cm of the commercial bentonite) and filtrate volume of 14mL/30min (which is less than the API recommended maximum filtrate loss of 15mL/30min).

Conclusions

This work presents a new lignocellulose-based additive with a unique property that makes it a good candidate for facilitating filter cake removal, mitigate formation damage, and make drilling fluid less toxic. The additive can be made from cheap locally available resources. It could, therefore, serve to minimize the oil production costs that are associated with drilling fluids.

Integrated Green Pretreatment Methods for Oil Palm Lignocellulosic Biomass Valorization into Biofuels and Platform Chemicals

Conference: International Congress on Sustainability Science Engineering ICOSSE

Year: 2019

Proceeding: International Congress on Sustainability Science & Engineering (ICOSSE '19)

Group: Abstract Submissions

Session: Session 5 - Biomass conversion and utilisation II

Time: Tuesday, July 2, 2019 - 5:10pm-5:35pm

Authors:

Abdul Rahman, A. Z., Universiti Teknologi Malaysia
Saidina Amin, N. A., University Technology Malaysia

Lignocellulosic biomass is a by-product of oil palm industry, which is abundantly available in Malaysia.
Upgrading this biomass into platform chemicals and biofuels will lead to an increase in the value of
the oil palm industry and contribute to the national economy. A number of novel pretreatment methods has been devised to degrade the structure of the lignocellulosic composite that makes up
the plant biomass material. Ionic liquid pretreatment is a green solvent capable of dissolving cellulose,
but it suffers from high cost. Deep eutectic solvent has similar properties to ionic liquid, capable of
dissolving lignin, with the added benefit of being cheaper than most ionic liquids. In this study, a series
of green pretreatment methods are integrated in order to chemically valorize oil palm empty fruit
bunch to produce platform chemicals and biofuels, namely cellulose nanocrystal, levulinic acid and
ethyl levulinate. Optimizations and kinetic studies will contribute towards potentially commercializing
this process at industrial scale.

Numerical Simulation of Lignocellulosic Biomass Ozonolysis in a Fixed - Bed Reactor

Conference: International Congress on Sustainability Science Engineering ICOSSE

Year: 2019

Proceeding: International Congress on Sustainability Science & Engineering (ICOSSE '19)

Group: Abstract Submissions

Session: Session 5 - Biomass conversion and utilisation II

Time: Tuesday, July 2, 2019 - 5:35pm-6:00pm

Authors:

Shamjuddin, A., Universiti Teknologi Malaysia
Saidina Amin, N. A., Universiti Teknologi Malaysia
Ab Rasid, N. S., Universiti Teknologi Malaysia

Oil palm waste (OPW) is known as one of the abundant agricultural wastes, which has high lignocellulose content. Lignocellulosic biomass is one of the renewable sources of energy stored in green plants and other organic matter. This research aims to enhance bioenergy production by pre-treating the lignocellulosic biomass particularly the OPW with zero waste in a sustainable process. This paper proposes ozone pre-treatment as a substantial potential and promising pre-treatment method, for renewable energy production in Malaysia. This study highlights the utilization of COMSOL finite element software in developing a coupled numerical model to simulate the velocity of ozone and to determine the concentration profile of lignin in the reactor. The flow behaviours of ozone in the free and porous-bed reactor regions were investigated. The flow behaviour of ozone during ozonolysis process depended on the permeability and porosity of biomass feedstock. The results from this study could be extended to include biomass from other sources such as wood waste. Thus, the findings from this research will have potentials in advancing technology innovation for enhancing resource sustainability by producing biofuels.

Keywords: Ozonolysis, COMSOL, Flow, Lignocellulosic biomass, Transport, Numerical simulation

Functionalized Hollow Mesoporous Silica Sphere with Magnesium Bromide for Ammonia Absorption Enhancement

Conference: International Congress on Sustainability Science Engineering ICOSSE

Year: 2019

Proceeding: International Congress on Sustainability Science & Engineering (ICOSSE '19)

Group: Abstract Submissions

Session: Session 6 - Waste handling, management and minimization II

Time: Wednesday, July 3, 2019 - 9:15am-9:40am

Authors:

Ming, Z., NJtech
Hua, M., NJtech
Xingxiang, P., NJtech

Ammonia is hazardous pollutant that trace amount of this chemical can be toxic to human respiratory system. Traditional method by using porous material like active carbon or alumina is less efficient in ammonia adsorption. Although magnesium bromide ($MgBr_2$) can absorb ammonia to form chemical complex, the poverty of pores make it less likely to achieve its theoretical capacity. In order to overcome shortcomings of traditional materials, we impregnate magnesium bromide into hollow mesooporous silica sphere (HMSS). The HMSS is synthesized via template method with polystyrene nanoparticle (PSP) and cetyltrimethylammonium bromide (CTAB) as structural precursors. HMSS of diameter 200nm and 400nm are synthesized respectively by controlling chain length of PSP during polymerization. Experimental results show that magnesium bromide impregnated on 200nm HMSS can achieve higher amount of ammonia capacity. It indicates a promising material for enhanced ammonia absorption.

Fabrication of Novel Concrete Blocks By Merging Ancient and Modern Techniques

Conference: International Congress on Sustainability Science Engineering ICOSSE

Year: 2019

Proceeding: International Congress on Sustainability Science & Engineering (ICOSSE '19)

Group: Abstract Submissions

Session: Session 6 - Waste handling, management and minimization II

Time: Wednesday, July 3, 2019 - 9:40am-10:05am

Authors:

Shukla, A., Indian Institute of Technology Delhi
Bhaskarwar, A. N., Indian Institute of Technology Delhi

The demand for green buildings is growing amongst rising global-level environmental concerns. Traditionally, lime was used in India as a binder. It is more durable than cement, but the strength is less. In order to improve the strength of lime based binder, many combinations of additives such as jaggery, molasses, curd, and egg yolk were used in the past, and the resultant mortar was called as lime mortar. There are many examples of the durability of lime-mortar based structures such as Mohanjo-Daro, which is a 4000-years old heritage monument of the ancient Indian civilization. Powdered burnt-brick material, termed as Surkhi, was used as â €˜the aggregateâ€™ along with jaggery.

Here, we report a combination of the ancient technique of jaggery and lime with the modern day problem of waste utilization, in which fly ash was used as the supplementary cementitious material. Jaggery solutions were prepared at different concentrations and temperatures along with lime in different ratios to set the matrix. Jaggery and sugar were both used along with lime as an additive to make pre-fabricated fly-ash blocks. These pre-fabricated blocks are environment-friendly, since these compositions can reduce the net CO_2 emission and also simultaneously utilize the fly ash, provided we wisely design the capture of CO_2 at source itself while producing lime from the limestone. Large surface areas of long-lasting buildings can provide an effective sink for the capture of CO_2 from the usual low atmospheric concentrations over centuries of such buildingsâ€™ lives.

An Innovative Way to Recycle Glass in Hong Kong

Conference: International Congress on Sustainability Science Engineering ICOSSE

Year: 2019

Proceeding: International Congress on Sustainability Science & Engineering (ICOSSE '19)

Group: Abstract Submissions

Session: Session 6 - Waste handling, management and minimization II

Time: Wednesday, July 3, 2019 - 10:05am-10:30am

Authors:

Bhardwaj, A. A., Columbia University

Hong Kong is a bustling international hub with a glaring waste issue. The small island has already filled 13 of its 16 initial landfills, and the remaining three will reach capacity within a few years. The Hong Kong government hopes to improve waste diversion from landfill by bolstering the cityâ€™s recycling sector. However, glass is not included in the set of materials recycled through the cityâ€™s collection system. As a result, 93% of glass, mostly bottles, is disposed instead of recycled in Hong Kong, adding over 300 tonnes of glass to the landfills daily. However, the capacity for recycling glass is readily availableâ€"the Tiostone Environmental recycling plant has developed a novel use for glass. The plant crushes glass and uses it to replace river sand in the process to manufacture paving bricks. Unsurprisingly, the plant operates far below capacity because so little glass is recycled in the city. Due to the governmentâ€™s inaction in creating a stream of recycled glass, I founded Glassroots, a Non-Profit aimed at increasing the amount of glass sent to Tiostone for recycling. I worked closely with Tiostone to establish and promote glass recycling collection points throughout the city, building a grassroots movement behind recycling glass. In this presentation, I will detail the chemical process used to recycle glass into bricks, and contextualize this process in the socioenvironmental state of Hong Kong.

Heavy Metals Toxicity Is Reduced during Valorization of Biowastes through Vermicomposting

Conference: International Congress on Sustainability Science Engineering ICOSSE

Year: 2019

Proceeding: International Congress on Sustainability Science & Engineering (ICOSSE '19)

Group: Abstract Submissions

Session: Session 6 - Waste handling, management and minimization II

Time: Wednesday, July 3, 2019 - 10:30am-10:55am

Authors:

Khan, M. B., Ministry of Education Key Laboratory of Environmental Remediation and Ecological Health
Yang, X., Ministry of Education Key Laboratory of Environmental Remediation and Ecological Health
Shijun, Z., Ministry of Education Key Laboratory of Environmental Remediation and Ecological Health, College of Environmental and Resource Sciences, Zhejiang University, Hangzhou 310058, China.

Sewage sludge and kitchen refuse are ubiquitously mounting wastes with high organic load, which if reprocessed they could salvage the environment. Reckoned with this certitude, an incubating study was initiated on sequential preincubation of sewage sludge with kitchen waste in 100:0, 70:30, 50:50, and 30:70 ratios ensued by vermicomposting using *Eisenia fetida*. Earthworm growth parameters data reflected that sewage sludge and kitchen waste mixture with 70:30 ratio increased the number of cocoons (10.6%), biomass (8.2%), growth rate (8.3%), reproduction rate (12.2%), and decreased their mean mortality rate (80.1%) as compared to that in sole sewage sludge (control). Results of chemical analysis and SEM/EDS imaging, showed that alkalinity, organic carbon, C/N ratio, organic matter and concentration of trace elements (Cd, Cr, Cu, Mn, Pb and Zn) reduced while macronutrients (N, P, K, Ca and Mg) increased in the final vermicompost as compared to that in initial mixtures. The FT-IR analysis also revealed that various biochemical functional groups underwent biodegradation during combined preincubation-vermicomposting. Bioaccumulation factor (BAF) of all trace elements in the earthworm tissues was higher with 70:30 ratio of substrates, with the trend of Cd > Zn > Cu > Mn > Pb > Cr. Hence, this study concludes that combined preincubation-vermicomposting is the most efficient and ecofriendly technique for biodegradation, stabilization and conversion of sewage sludge and kitchen waste into organic fertilizer. The nutrient rich vermicompost can be safely used as horticultural substrate and soil conditioner for efficient management of degraded soils. Finally, combined preincubation-vermicomposting is a sustainable system of recycling the sewage sludge along with kitchen waste.

Building Sustainable Biosafety & Biosecurity for Synthetic Biopharmaceuticals

Conference: International Congress on Sustainability Science Engineering ICOSSE

Year: 2019

Proceeding: International Congress on Sustainability Science & Engineering (ICOSSE '19)

Group: Abstract Submissions

Session: Session 7 - Sustainable technologies development II

Time: Wednesday, July 3, 2019 - 10:55am-11:20am

Authors:

Jyothikumar, V., Syngene International Limited

International prevention and control of infectious diseases requires a robust diseases surveillance and diagnostics platform. The major biopharmaceutical concern, particularly with microbial systems, is that genetically modified self-replicating cells may produce gene products with unknown consequences if there is a loss of containment to the environment. To address this biosafety concern, many strategies considered to address the mechanisms for constraining microbial cell replication and horizontal gene transfer. Recombinant containment strategies are considered to built into vector systems like preventing the spread of recombinant and synthetic DNA to host organisms either through horizontal gene transfer (HGT) or avoiding loss of containment of the engineered organism releases to environment. However, synthetic biology biosafety also depends on the genetic cargo of the vector and the protein with an unknown pharmacological or environmental effect. The explosions of interests in biosafety and biosecurity in recent times with steady increase in facilities and labs handling pathogens, nanoparticles, markers, toxins, biomaterials for 3D printing in translational research, cell therapy and personalized medicines have made biosafety-biosecurity awareness and knowledge is more important than ever. Biosafety is about applying right knowledge, techniques and equipmentâ€™s to prevent exposures to potentially infectious materials or biohazards. Biosecurity is the process employed for ensuring biological agents are properly secured and protected against theft, or unauthorized access or use/release. One of the most important decisions to be made in working with biohazardous agents is the selection or development of work practices to prevent exposures. An agent- and protocol-based risk assessment should provide sufficient information to allow for the selection of the appropriate level of containment for the work. Importantly, security in biology differs from that in a nuclear or chemical enterpriseâ€”biological materials are ubiquitous in nature and due its covert nature could induce an epidemic with less time to react and contain the exposure.

Effect of Diesel on the Performance of Microemulsion Fuels

Conference: International Congress on Sustainability Science Engineering ICOSSE

Year: 2019

Proceeding: International Congress on Sustainability Science & Engineering (ICOSSE '19)

Group: Abstract Submissions

Session: Session 7 - Sustainable technologies development II

Time: Wednesday, July 3, 2019 - 11:20am-11:45am

Authors:

Abrar, I., Indian Institute of Technology Delhi
Bhaskarwar, A. N., Indian Institute of Technology Delhi

The increased consumption and limited reserves of diesel, along with its contribution to emissions of particulate matter and nitrogen oxides, has led to an interest in the diesel alternatives, such as biodiesel, hydrotreated vegetable oils, emulsions, and microemulsions. Microemulsions are thermodynamically stable and optically isotropic dispersions of oil and water, stabilized by layers of surfactant and/or co-surfactant. They do not involve any reactions during preparation, and can be used directly in the engine without any modifications. The current work examines the usability of diesel-based microemulsions as an alternative fuel for the compression-ignition engines. Different compositions of microemulsions were formulated with 35 and 40 %(v/v) alcohol, 57.8 â€' 64.4 %(v/v) diesel, and 0.6 â€' 2.2 %(v/v) water, and their properties tested as per ASTM D975. Most of the microemulsion compositions satisfied the ASTM standards, and therefore were further tested in a diesel engine to evaluate the performance and emissions parameters. The engine studies were carried out with microemulsions having constant alcohol percentages, and varying diesel percentages. It was observed that almost all the microemulsions had slightly higher fuel consumption, due to a reduction in calorific value. However, the thermal efficiencies were slightly improved due to the presence of alcohol and water, leading to a more complete combustion of the fuel. The CO emissions were found to be higher at lower loads, and lower at higher loads, when compared to those of the neat diesel. The emissions of CO_2, unburnt HC, and of NO_x were lower throughout all the loads for most of the microemulsions. These microemulsions were also found to be economical as compared to petro-diesel in the present Indian scenario. The microemulsions could successfully replace almost 40% of diesel, with lesser emissions, and without any additional modifications in the engine, and thus are a promising cleaner alternative to diesel.

A Novel Semi-Automatic Method to Optimize Multi-Lamp High Flux Solar Simulators

Conference: International Congress on Sustainability Science Engineering ICOSSE

Year: 2019

Proceeding: International Congress on Sustainability Science & Engineering (ICOSSE '19)

Group: Abstract Submissions

Session: Session 7 - Sustainable technologies development II

Time: Wednesday, July 3, 2019 - 11:45am-12:10pm

Authors:

Mohamed Ali, A., Texas A&M University at Qatar
Hafeez, S., Texas A&M at Qatar
Hassan, M., Texas A&M at Qatar
Kakosimos, K. E., Texas A&M University at Qatar

Extensive research is being conducted on photon energy driven reactions which include wastewater treatment, H2 production, and advanced material aging under diffuse or concentrated light. Solar simulators are a convenient tool for the latter research applications because they allow emulation of solar radiation under well controlled laboratory conditions. Most concentrated light simulators are comprised of multiple light sources with ellipsoidal /parabolic mirrors. Thus, they require accurate characterization of various system parameters such as peak flux and flux density distribution. They also require optimization to ensure the system can operate at its theoretical maximum flux and provide the necessary radiating energy. However, this process is either manual or semi-automated and demands tedious expert intervention. This study utilizes the High Flux Solar Simulator (HFSS) facility at TAMUQ comprised of seven short-arc xenon lamps of 6 kW each. We present intermediate results of our attempts to create an automated system used to optimize alignment of the light sources (four degrees of freedom each) given variable system parameters. An automated data acquisition program was developed which included image capture, target movement, flux readings, etc. The irradiance was then characterized using the flux mapping method using an in-house algorithm used to calculate flux parameters. Monte-Carlo ray tracing was used to generate training data. This data was used to train a supervised machine learning algorithm based on a typical convolutional neural network model. Finally the model was tested against the same Monte-Carlo ray tracing used initially as well as experimental data from the HFSS. The proposed methodology is expected to facilitate initial deployment of HFSS. It will also assist on the dynamic control of reactor conditions i.e. emulating variable overcast or daily sunlight variability.

P-Graph Approach for Optimization of Segregated Targeting Networks

Conference: International Congress on Sustainability Science Engineering ICOSSE

Year: 2019

Proceeding: International Congress on Sustainability Science & Engineering (ICOSSE '19)

Group: Abstract Submissions

Session: Session 7 - Sustainable technologies development II

Time: Wednesday, July 3, 2019 - 12:10pm-12:35pm

Authors:

Aviso, K., De La Salle University
Jain, S., Indian Institute of Technology Bombay
Pimentel, J., Universidad Nacional de Colombia
Tan, R., De La Salle University
Bandyopadhyay, S., Indian Institute of Technology Bombay
Friedler, F., Pazmany Peter Catholic University

Segregated targeting networks are a special class of source-sink networks, addressed in the literature using Pinch Analysis technique. This work presents a novel approach for the optimization of segregating targeting problems using P-graph methodology. Along with the general constraints of satisfying the requirement of different demands (flow and material balance), this work also considers many practical constraints which are not included earlier for segregated targeting networks. For example, transportation cost associated with the flow of materials from sources to sinks, investment and fixed cost associated with different sources and sinks, etc. Apart from the inclusion of these practical constraints, P-graph methodology provides the advantage of generating n-best solution networks. This approach of addressing segregated targeting problems is demonstrated by using an illustrative case study from carbon constraint energy sector planning domain, which includes the optimal allocation of different energy sources to meet the demands of various sectors of the society.

Multi-Objective Sustainable Supply Chain Optimization of the Petroleum Industry in Kuwait

Conference: International Congress on Sustainability Science Engineering ICOSSE

Year: 2019

Proceeding: International Congress on Sustainability Science & Engineering (ICOSSE '19)

Group: Abstract Submissions

Session: Session 7 - Sustainable technologies development II

Time: Wednesday, July 3, 2019 - 12:35pm-1:00pm

Authors:

Elghool, A., Kuwait University

Multi-objective Sustainable Supply Chain Optimization of the Petroleum Industry in Kuwait

Aya. M. H. Elghool[a], Ghanima Al-Sharrah[b], Haitham M. S. Lababidi[c]

Chemical Engineering Department, College of Engineering & Petroleum, P.O. Box 5969, Safat â €“ 13060 KUWAIT

[a]Email: aya.elghool@grad.ku.edu.kw

[b]Email: g.sharrah@ku.edu.kw

[c]Email: haitham.lababidi@ku.edu.kw

Most of the supply chain models, especially for an industry like petroleum, are based mainly on economic considerations, where cost and profit are the direct focus of the study, however, both environmental and social considerations have been drawing the attention of researchers and different organizations around the world as well. The main objective of this work is to develop a multi-objective model that considers all three aspects of sustainability (Economics, Environment, and Social) and implement it for the supply chain network of a petroleum organization in Kuwait. The model uses cost, IRCHS (Indiana Relative Chemicals Hazard Score), and the total number of accidents together with training hours per employee as economic, environmental, and social indicators respectively. GAMS software is used to optimize the model by jointly minimizing total production, logistics, lost demand and backlog penalties cost, as well as the organizationâ€™s emissions and major accidents, while maximizing the number of training hours per employee. Preliminary results showed that the contribution of the three objective functions of the model (i.e; economics, environmental, and social) is crucial for smooth operation and continuous production in the facilities of the supply chain.

Functional Characterization of a Novel Class of Gemini (Dimeric) Surfactants Synthesized from Sunflower Oil

Conference: International Congress on Sustainability Science Engineering ICOSSE

Year: 2019

Proceeding: International Congress on Sustainability Science & Engineering (ICOSSE '19)

Group: Abstract Submissions

Session: Session 7 - Sustainable technologies development II

Time: Wednesday, July 3, 2019 - 1:00pm-1:25pm

Authors:

Pal, N., Indian Institute of Technology (Indian School of Mines) Dhanbad
Mandal, A., Indian Institute of Technology (Indian School of Mines) Dhanbad

The field of natural resource-derived products is gaining significant interest in the current global energy scenario. Natural surfactants or â€œsurface-active agentsâ€ with favorable attributes find usability in myriad sectors comprising foaming and cleaning, corrosion inhibition, gelation, oil recovery, waste water treatment, biomedical research, textile processing and preparative chemistry. Growing need for indigeneity, ease of availability/manufacturability and cost-effectiveness, combined with novelty in â€œgreenâ€ surfactant science is a pivotal area of concern. In present study, a novel class of self-assembling gemini surfactants with unique â€œdimer structureâ€ were synthesized from plant-derived raw material, sunflower oil. Sunflower seed oil is rich in mono-unsaturated (oleic acid) and poly-unsaturated (linoleic acid) components, with small amounts of saturated fats. Non-ionic bis(monoglyceride-1-hydroxymethyl-2-fattyacidester)-Î±,Ï‰-alkanediether surfactants with four and six methyl spacers were synthesized via a stepwise process: transesterification, followed by two stages of nucleophilic substitution (S_N2) reaction. The synthesized surfactants were characterized by FT-IR, ^{1}H-NMR and TGA studies. Surfactants retained their structural integrity under high temperature conditions. The surfactant molecules exhibited self-aggregation ability at low concentrations, resulting in low critical micelle concentration (CMC). Favorable interfacial behavior was observed at oil-aqueous interfaces. Addition of salt showed synergistic effects, resulting in ultralow interfacial tension (IFT) of the order of 10â ˆ'2 to 10â ˆ'3 mN/m. The gemini surfactant solution exhibited non-Arrhenius, pseudoplastic flow behavior, wherein viscosity increased with temperature due to greater degree of entanglement of worm-like micelles/aggregates to develop network structure in bulk solution. These formulated surfactants also formed stable oil-in-water emulsions with slow rate of oil droplet coaleascence, as observed from microscopic imaging. The surfactants, therefore, exhibited beneficial properties in terms of thermal stability, interfacial activity, viscosity behavior and emulsification. The sunflower oil based gemini surfactants, prepared from readily available and economical raw material, were successfully formulated as a promising alternative to synthetic, non-biodegradable surfactants that are commercially employed in industrial applications.

AUTHOR INDEX

AIChE
120 Wall Street, FL 23
New York, NY 10005-4020

CURRAN ASSOCIATES INC.
proceedings.com

ISBN 978-1-7138-0567-0